Splendor Solis

1582

Alchemical Wanderings

1598

Salomon Trismosin

Introduction by

J.K. (1920)

When I was a young fellow, I came to a Miner named Flocker, who was also an Alchemist, but he kept his knowledge secret, and I could get nothing out of him. He used a Process with common Lead, adding to it a peculiar Sulphur, or Brimstone, he fixed the Lead until it became hard, then fluid, and later on soft like Wax.

Of this prepared Lead, he took 20 Loth (10 ounces), and 1 mark pure unalloyed Silver, put both materials in flux and kept the composition in fusion for half an hour. Thereupon he parted the Silver, cast it in an ingot, when half of it was Gold.

I was grieved at heart that I could not have this art, but he refused to tell his secret process.

Shortly thereafter he tumbled down a mine and no one could tell what was the artifice he had used.

As I had seen it really done by this miner, I started in the year 1473 on my travels to search out an artist in Alchemy, and where I heard of one I went to him, and in these wanderings I passed 18 months, learning all kinds of Alchemical Operations, of no great importance, but I saw the reality of some of the particular processes, and I spent 200 Florins of my own money, nevertheless I would not give up the search. I thought of boarding with some of my friends, and took a journey to Laibach, thence to Milan, and came to a monastery. There I heard some excellent lectures and served as an assistant, for about a year.

Then I travelled about, up and down in Italy, and came to an Italian tradesman, and a Jew, who understood German. These two made English Tin look like the best fine Silver, and sold it largely. I offered to serve them. The Jew persuaded the Trader to take me as a Servant, and I had to attend the fire, when they operated with their art I was diligent, and they kept nothing from me, as I pleased them well. In this way I learnt their art, which worked with corrosive and poisonous materials, and I stopped with them fourteen weeks.

Then I journeyed with the Jew to Venice. There he sold to a Turkish merchant forty pounds of this Silver. While he was haggling with the merchant I took six Loth of the Silver, and brought it to a Goldsmith, who spoke Latin, and kept two Journeymen, and I asked him to test the Silver. He directed me to an Assayer on Saint Marks' Place, who was portly and wealthy. He had three German Assay-assistants. They soon brought the Silver to the test with strong acids, and refined it on the Cupel; but it did not stand the test, and all flew away in the fire. And they spoke harshly to me asking where I got the Silver. I told them I had come on purpose to have it tested, that I might know if it was real silver.

When I saw the fraud, I returned not to the Jew, and paid no more attention to their art, for I feared to get into trouble together with the Jew, through the false silver.

I then went to a College in Venice, and asked there if they could give me two meals daily while I looked for employment. The Rector told me of a Hospital where there were other Germans, and there we got sumptuous food. It was an Institution for destitute strangers, and people of all nations came there.

The next day I went to Saint Marks' Place, and one of the Assay assistants came up, and asked me where I got that Silver? Why I had it tested, and if I had any more of it? I said I had no more of that silver, and that I was glad to have got rid of it, but I had the art and I should not mind telling it to him. That pleased the Assayer, and he asked me if I could work in a Laboratory? I told him I was a Laborant travelling on purpose to work in alchemical Laboratories. That pleased him vastly, and he told me of a nobleman who kept a laboratory, and who wanted a German Assistant. I readily accepted, and he took me straight to the Chief Chemist, named Tauler, a German, and he was glad to get me. So he engaged me on the spot at a weekly wage of two crowns and board as well. He took me about six Italian miles out of Venice to a fine large mansion called Ponteleone. I never saw such Laboratory work, in all kinds of Particular Processes, and medicines, as in that place. There everything one could think of was provided and ready for use. Each workman had his own private room, and there was a special cook for the whole staff of Laboratory assistants.

The Chief Chemist gave me at once an Ore to work on, which had been sent to the nobleman, four days previously. It was a Cinnabar the Chief had covered with all kinds of dirt, just to try my knowledge, and he told me to get it done within two days. I was kept busy, but succeeded with the Particular Process, and on testing the ingot of the fixed Mercury, the whole weighed nine Loth, the test gave three Loth fine Gold.

That was my first work and stroke of luck. The Chief Chemist reported it to the nobleman, who came out unexpectedly, spoke to me in Latin, called me his Fortunatum, tapped me on the shoulder and gave me twenty-nine crowns. He spoke a funny kind of Latin I could hardly understand, but I was pleased with the money.

I was then put on oath not to reveal my Art to anyone. To make a long story short, everything had to be kept secret, as it should be. If someone boasts of his art, even if he has got the Truth, God's Justice will not let such a one go on. Therefore be silent, even if you have the highest Tincture, but give Charity.

I saw all kinds of operations at this Nobleman's Laboratory, and as the Chief Chemist favoured me, he gave me all kinds of operations to do, and also mentioned, that our employer spent about 30,000 Crowns on these arts, paying cash for all manner of books in various languages, to which he gave great attention. I myself witnessed that he paid 6,000 Crowns for the Manuscript Sarlamethon. A process for a Tincture in the Greek Language. This the nobleman had soon translated and gave me to work. I brought that process to a finish in fifteen weeks. Therewith I tinged three metals into fine Gold; and this was kept most secret. This nobleman was gorgeous and powerful, and when once a year the Signoria went out to sea, to witness the throwing of a Gem Ring into the water at the ceremony of wedding the Adriatic, our gentleman with many others of the Venetian nobility went out in his grand pleasure ship, when suddenly a hurricane arose and he with many others of the Venetian Lords and Rulers, was drowned.

The Laboratory was then shut up by the family, the men paid off, but they kept the Chief Chemist.

Then I went away from Venice, to a still better place for my purpose, where Cabalistic and Magical books in Egyptian language were entrusted to my care, these I had carefully translated into Greek, and then again retranslated into Latin. There I found and captured the the Treasure of the Egyptians. I also saw what was the great Subject they worked with, and the ancient Heathen Kings used such Tinctures and have themselves operated with them, namely, Kings Xofar, Sunsfor, Xogar, Xophalat, Julaton, Xoman and others. All these had the great treasures of the Tincture and it is surprising that God should have revealed such Secrets to the Heathen, but they kept it very secret.

After a while I saw the fundamental principles of this art, then I began working out the Best Tincture (but they all proceed, in a most indescribable manner from the same root), when I came to end of the Work I found such a beautiful red colour as no scarlet can compare with, and such a treasure as words cannot tell, and which can be infinitely augmented. One part tinged 1,500 pages Silver into Gold. I will not tell how after manifold augmentation what quantities of Silver and other metal I tinged after the Multiplication. I was amazed.

Study what thou art,
Whereof thou art a part,
What thou knowest of this art,
This is really what thou art.
All that is without thee
Also is within,
Thus wrote Trismosin.

Arma Artis

Eamus
Quesitum Quatuor
Elementorum
naturas

Particularia
Via Universalis particularibus Inclusa

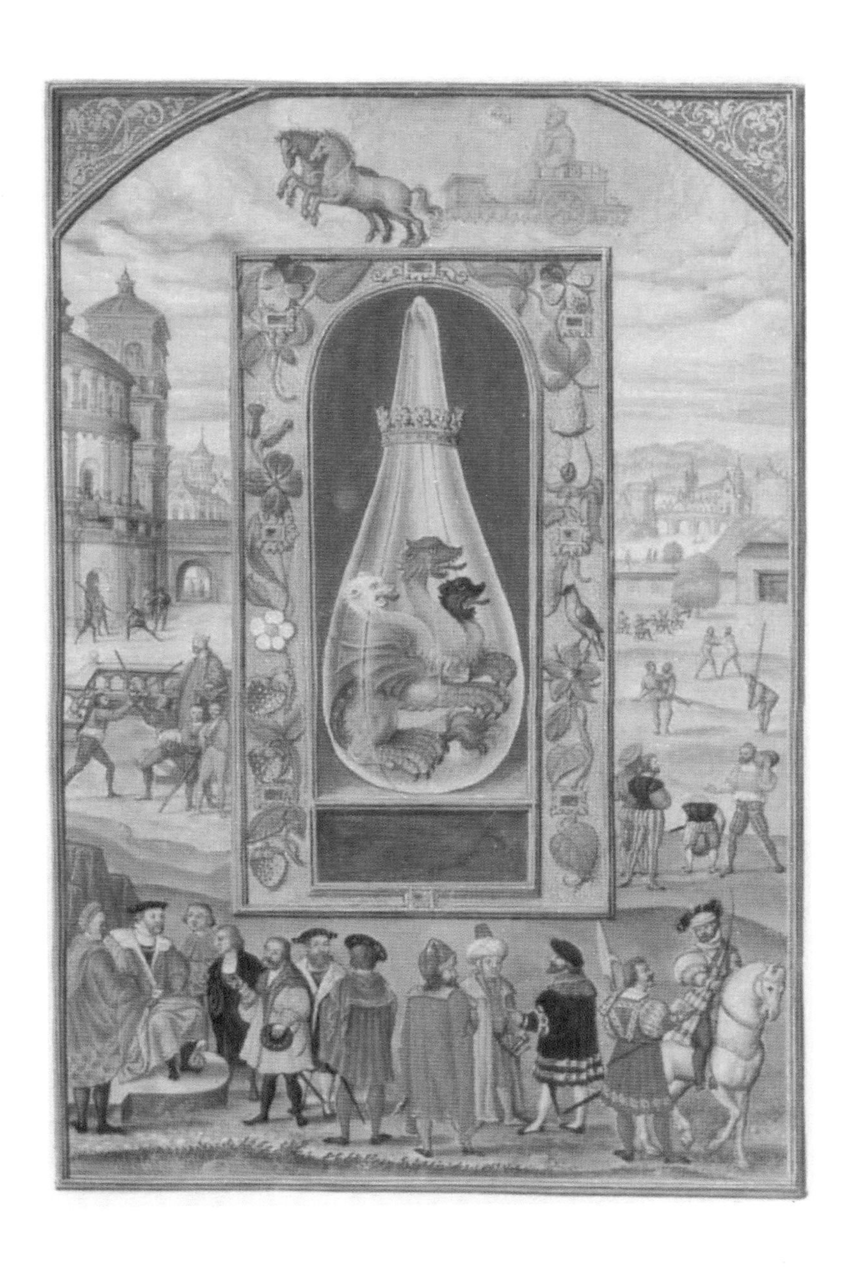

Plate I. – SPLENDOR SOLIS - The Arms of the art

**THIS BOOK
IS NAMED
SPLENDOR SOLIS
OR
Splendour of the Sun**

AND IS DIVIDED INTO SEVEN PARTS, IN WHICH IS DESCRIBED THE HIDDEN MYSTERY OF THE OLD PHILOSOPHERS, AS WELL AS ALL THAT NATURE REQUIRES TO CLEARLY ACCOMPLISH THE WHOLE WORK, INCLUDING ALL THE ADDED THINGS; AFTER WHICH NO ONE SHALL BE ADVISED

TO GRAPPLE WITH THE MYSTERY OF THE NOBLE ART WITH HIS OWN SENSES.

ALPHIDIUS, one of the old Philosophers, said: "Every one who does not care for the trouble of obtaining the Philosopher's Stone, will do better in making no enquiries at all than only useless ones."

The same also says RHASES, in his book "LIGHT OF LIGHTS": "Let it be said then to all, I hereby admonish them most earnestly, that none be so foolhardy to presume to understand the unknown intermixture of the elements, for as ROSINUS say: "All who engage in this Art, and are wanting the knowledge and perception of things, which the Philosophers have described in their books, are erring immensely; for the Philosophers have founded this art in a natural beginning, but of a very hidden operation. Though it is evident that all corporeal things originate in and are maintained and exist of the Earth, according to Time and Influence of the Stars and Planets, as: Sun, Moon and the others, together with the four qualities of the elements, which are without intermission, moving and, working therein, thereby creating every growing and procreating thing in its individual form, sex and substance, as first created at the Beginning by God, the Creator, consequently all metals, originate in the earth of a special and peculiar matter produced by the four properties of the four elements, which generate in their mixture the metallic force, under the influence of their respective planets.

All this is well described by the natural master ARISTOTLE, in the fourth Book METEOROLOGICORUM, when he says, that QUICKSILVER is a matter common to all metals. But it must be known that first in Nature is the compound matter of the four elements.

In acknowledging this property of Nature, the Philosophers called their Matter MERCURIUS, or QUICKSILVER. How this MERCURIUS takes the perfect form of Gold, silver or other metals through the working of nature need not be mentioned here. The teachers of Nature's Philosophy describe it sufficiently in their books.

Upon such is based and founded the ART of the Philosopher's Stone; for it originates in Nature, thence follows a natural end in a just form, through just and natural means.

Plate II – THE FIRST TREATISE - Philosopher with flask

THE FIRST TREATISE

IN THE FOLLOWING TREATISE WE SHALL DISCOURSE ON THE ORIGIN OF THE STONE OF THE PHILOSOPHERS AND THE ART HOW TO PRODUCE IT.

THE Philosopher's Stone is produced by means of the Greening and Growing Nature.

HALI the Philosopher, says there of: "This Stone rises in growing, greening things." Wherefore when the Green is reduced to its former nature, whereby things sprout and come forth in ordained time, it must be decoded and putrefied in the way of our secret art. That by Art may be aided, what Nature decocts and putrefies, until she gives it, in due time, the proper form, and our Art but adapts and prepares the Matter as becomes Nature, for such work, and for such work provides also, with premeditated Wisdom, a suitable vessel.

For Art does not undertake to produce Gold and Silver, anew, as it cannot endow matter with its first origin, nor is it necessary to search our Art in the places and caverns of the earth, where minerals have their first beginning. Art goes quite another way to work and with different intention from Nature, therefore does Art also use different tools and instruments.

For that reason can Art produce extraordinary things out of the aforesaid natural beginnings such as Nature of herself would never be able to create. For unaided Nature does not produce things whereby imperfect metals can in a moment be made perfect, but by the secrets of Our Art this can be done. Here Nature serves Art with Matter, and Art serves Nature with suitable Instruments and method convenient for Nature to produce such new forms; and although the before mentioned Stone can only be brought to its proper form by Art, yet the form is from Nature. For the form of every thing be it living, growing, or metallic, comes into existence by virtue of the interior force in matter—except the human soul.

But it must be borne in mind that the essential form cannot originate in matter unless it is by the effect of an accidental form, not by virtue of that form, but by virtue of another real substance, which is the Fire or some other accidental active heat. By way of allegory, we take a hen's egg; in this the form of the chicken can not take shape, without the presence and aid of accidental form, which is the intermixture of the red with the white, by virtue of the heat coming from the hatching hen, and although the egg is the hen's material, nevertheless it cannot develop either its real or accidental form, otherwise than by putrefaction, which is caused by the influence of warmth, so can also neither the real nor the accidental form of the Philosopher's Stone originate in their natural matter without the agency of

Putrefaction or Decoction, of which we shall speak hereafter. PUTREFACTION takes place when the natural heat of a moist body is expelled by an external heat, or else when the natural heat of the subject is destroyed by cold. For then the natural warmth leaves everything and gives room to putrefaction. The Philosophers do not mean this kind of Putrefaction. Their Putrefaction is a moistening of dry bodies, that they may be restored to their former state of Greening and Growing. In this process of Putrefaction, moist and dry are joined together and not destroyed, but the moisture is quite separated from the dryness, then it is necessary to separate the dry parts that turned to ashes.

This Incineration the Philosophers will also not have, but they will have their Putrefaction, which is a drying, trituration and calcination, to be done in such wise, that the natural moisture and dryness be united together, but sepanted and dried up from the superfluous moisture that is destructive.

Even as the food is being absorbed on entering an animal's stomach, that it may be digested and changed and afterwards supply the feeding force and moisture necessary to the existence and augmentation of nature, and be separated of its superfluous parts. How then everything has to be fed in its way according to its nature will be shown in the aforesaid Philosopher's Stone.

Plate III – THE SECOND TREATISE - The Knight on the double fountain

THE SECOND TREATISE

MATTER AND NATURE OF

THE PHILOSOPHER'S STONE

MORIENUS says: You shall know that the whole work of this Art ends in two Operations hanging very close together, so that when the one is complete, the other may begin and finish, this perfecting the whole Mastery. But as they only act on their own matter, it is necessary to give more particulars about it. GEBER says in his "SUMMA PERFECTIONS MAGISTERI IN SUA NATURA" "that Nature produces the Metals from Mercury and Sulphur," and to the same effect we see FERRARIUS speak in his "TREATISE ON ALCHEMY," in the 25th chapter, that from the beginning of the Origin of

Metals, Nature also uses a slimy, heavy water, mixed with a very peculiar white sulphuric subtile earth, which resolves the former into a steam and vapour, raises it in the veins or crevices of the earth and decocts, steams and collects it together so long, till at last dryness and moisture completely unite, thereby forming the substance which we call Mercury, and which constitutes the peculiar and very first Matter of all metals, and again he treats of it in the 26th chapter as follows: "Those who will imitate nature, are not to use Mercury only, but Mercury mixed with Sulphur, but not the common Mercury and Sulphur, but those only which Nature herself has mixed, well prepared and decocted into a sweet fluid. In such a Mercury Nature has begun with primary action and ended in a metallic nature, having thus done her part, leaving the rest for Art to complete her work, into a perfect Philosopher's Stone.

From the aforesaid it will be seen that he who will proceed properly in this Art, shall according to all Philosophers, begin where nature has left off, and shall take that Sulphur and Mercury which nature has collected in its purest form, in which took place the immediate union, which otherwise cannot be accomplished by anybody without art.

In order to receive the force that penetrates such subtle Matter, some Alchemists calcinate Gold that they may dissolve it, and separate the elements until they reduce it to a volatile spirit or to the subtle nature of the greasy fumes of Mercury and Sulphur, and this then is the nearest matter, that combines most closely with gold, and receives the form of the occult Philosopher's Stone, this matter is called the Mercury of the Philosophers, about which ARISTOTLE, speaking to ALEXANDER the King, says : " Chose for our Stone that wherewith kings are decorated and crowned."

Though this Mercury alone is the matter and the one only thing and a combination of other things, yet is this thing so manifold in its effects, and in its names, that no one can find out the true meaning from the writings of the Philosophers, and this is done for the purpose as ROSINUS says, " that every one may not get at it."

It is at the same time a way of producing effects and a vessel wherein all things multiply themselves, because of the adjustment of all things comprised in Nature.

For now the Philosophers say: "dissolve the thing, and sublimate it, and then distil it, coagulate it, make it ascend, make it descend, soak it, dry it, and ever up to an indefinite number of operations, all of which take place at the same time and in the same vessel." ALPHIDIUS confirms this and says: "You must know that when we dissolve we sublimate as well and calcinate without interruption," and if our Corpus is being thrown into the water, for the purpose of dissolution, it first turns black, then separates itself, dissolving and sublimating, it unites itself with the spirit which is its origin and birth.

It has been compared as analogous to all things in the world, visible or invisible, possessed of a soul or not, corporeal or animal, dead or alive, mineral or vegetative; analogous to the elements and their compositions, to things hot and cold, further to all colours, all fruits, all birds, and in short to all things between Heaven and Earth, and among all these axe belonging to this Art the aforesaid operations, which are explained by the Philosophers in two word "Man and Wife," or "Milk and Cream." He who does not understand these does not understand the preparation of this Art.

Plate IV – THE THIRD TREATISE - The Solar King and Lunar Queen meet

THE THIRD TREATISE

NOW FOLLOWS THE MEANS WHEREBY THE WHOLE WORK OF THIS MASTERY IS PERFECTED; EXPLAINED BY A FEW SUITABLE ILLUSTRATIONS, PARABLES, AND VARIOUS APHORISMS OF THE PHILOSOPHERS

HERMES, a Father of Philosophy, says: "It is indeed needed that at the End of this World, Heaven and Earth should meet and come home".

Meaning by Heaven and Earth the aforesaid two Operations; but many doubts arise, before the Work is finished. That the following Figures may be better understood we give a few Parables in illustration.

Plate V. – THE FIRST PARABLE

AND THIS IS THE FIRST PARABLE

GOD created the Earth plain and coarse, and very productive of Gravel, Sand, Stones, Mountains and Valleys, but through the influence of the planets, and the working of Nature, the Earth has been changed into many forms. Outside there are hard stones, high mountains and deep valleys, and strange things and colours are inside the Earth, as, for instance, Ores and their beginnings, and with such things earth has come from the original form, in the following manner: Where the Earth first began to grow large, or to expand and multiply, the constant operation of the Sun-Heat also formed in the interior of the Earth a sulphury vapourous and damp heat, penetrating her through and through. This penetrating work of the Sun's heat caused in the cold and damp of the Earth, the formation of large quantities of vapour fumes, fog and gas, all of which grow with the length of time strong enough to follow their tendency to rise, thus causing on the Earth's surface eruptions, forming hill and dale, &c. Where there are such hills and dales, there the Earth has been matured and most perfectly mixed with heat and cold, moisture and dryness, and there the best ores may be found. But where the earth is flat there has been no accumulation of such fumes and vapours, and there no ores will be found, while the uplifted part of the soil, especially, such as has been slimy, loamy, and fat, and has been saturated with a moisture from on high; got soft again, forming dough-like layers one on top of the other, which in the course, of time, under the influence of the Sun's heat, become more and more firm, hard and baked; and other ground as gravel and sand, brittle and yet soft, hanging together like grapes, is too meagre and dry, and has not received enough moisture, consequently it could not form itself into layers, but remained full of holes, like badly prepared pap, or like a mealy dough, which has not been watered enough; for no earth can become stone, unless it be rich and slimy and well mixed with moisture. After the drying up of the water by the Sun's heat, the fat substance will keep the ground together, as otherwise it would remain brittle and fall to pieces again. That which has not become perfectly hard as yet, may become so, and turn to stone, under the constant influence of the Sun's heat and Nature, as well as the aforesaid fumes and gases originating in the properties of the elements, which are by these means still being operated upon in the interior of the earth, and when they seize upon watery vapours with a pure, subtle earthy substance, then they form the Philosophers' Mercury; but when they are solid and brought to a fiery, earthy and subtle hardness, then will the Philosophers' Sulphur be the result.

About this Sulphur HERMES says: "It will receive the powers of the highest and lowest planets, and with its force it penetrates solid things, it overcomes all matter and all precious stones."

Plate VI. – THE OTHER PARABLE

THE OTHER PARABLE

HERMES, the First Master of this Art, says as follows: "The Water of the Air, which is between Heaven and Earth, is the Life of everything; for by means of its Moisture and Warmth, it is the medium between the two opposites, as Fire and Water, and therefore it rains water on earth, Heaven has opened itself, and sent its Dew on earth, making as sweet as honey, and moist. Therefore the Earth flowers and bears manifold coloured blooms and fruits, and in her interior has grown a large Tree with a silver stem, stretching itself out to the earth's surface. On its branches have been sitting many kinds of birds, all departing at Daybreak, when the Ravenhead became white. The same tree bears three kinds of Fruit. The First are the very finest Pearls. The Second are called by Philosophers TERRA FOLIATA. The Third is the very purest Gold. This Tree gives us as well the fruit of Health, it makes warm what is cold, and what is cold it makes warm, what is dry it makes moist, and makes moist what is dry, and softens the hard, and hardens the soft, and is the end of the whole Art. Thereof says the Author of "The Three Words," "The Three Moistures are the most precious Words of the whole Mastery." And the same says GALENUS, when he speaks of the Herb LUNATICA or BERISSA. Its root is a Metallic Earth; it has a red stem, spotted with black, grows easily and decays easily, and gains Citrine Flowers after three days; if it is put in Mercury, it changes itself into perfect Silver, and this again by further decoction changes into Gold, which then turns hundred parts of Mercury into the finest Gold. Of this tree speaks VIRGILIUS, in the sixth book of the AENEIDE, when he relates a Fable, how AENEAS and SILVIUS went to a tree, which had golden branches, and as often as one broke a branche off, another one grew in its place.

Plate VII. – THE THIRD PARABLE

THE THIRD PARABLE

AVINCENA says in the Chapter on the MOISTURES:— "When Heat operates upon a moist body, then is blackness the first result." For that reason have the old Philosophers declared they saw a Fog rise, and pass over the whole face of the earth, they also saw the impetuosity of the Sea, and the streams over the face of the earth, and how the latter became foul and stinking in the darkness. They further saw the King of the Earth sink, and heard him cry out with eager voice: "Whoever saves me shall live and reign with me for ever in my brightness on my royal throne," and Night enveloped all things. The day after they saw over the King an apparent Morning Star, and the Light of Day clear up the darkness, the bright Sunlight pierce through the clouds, with manifold coloured rays, of brilliant brightness, and a sweet perfume from the earth, and the Sun shining clear. Herewith was completed the Time when the King of the Earth was released and renewed, well apparelled, and quite handsome, surprising with his beauty Sun and Moon. He was crowned with three costly crowns, the one of Iron, the other of Silver, and the third of pure Gold. They saw in his right hand a Sceptre with Seven Stars, all of which gave a Golden Splendor, and in his left hand a golden Apple, and seated upon it a white Dove, with Wings partly silvered and partly of a golden hue, which ARISTOTLE so well spoke of when he said: "The Destruction of one thing is the birth of another." Meaning in this Masterly Art: "Deprive the thing of its Destructive Moisture, and renew it with its own Essential one which will become its perfection and life."

Plate VIII. – THE FOURTH PARABLE

THE FOURTH PARABLE

MENALDUS the Philosopher, says: "I command all my descendants to spiritualise their bodies by DISSOLUTION, and again to materialise the spiritual things by means of a gentle decoction. Mentioning which SENIOR speaks thus: "The Spirit dissolves the body, and in the Dissolution extracts the Soul of the Body, and changes this body into Soul; and the Soul is changed into the Spirit, and the Spirit is again added to the Body, for thus it has stability." Here then the body becomes spiritual by force of the Spirit. This the Philosophers give to understand in the following Signature, or Figure: They saw a man black like a negro sticking fast in a black, dirty and foul smelling slime or clay; to his assistance came a young women, beautiful in countenance, and still more so in body, most handsomely adorned with many-coloured dresses, and she had wings on her back, the feathers of which were equal to those of the very finest white Peacock, and the quills were adorned with fine pearls, while the feathers reflected like golden mirrors. On her head she had a crown of pure gold, and on top of it a silver star around her neck she wore a necklace of fine Gold, with the most precious Ruby, which no king would be able to pay; her feet were clad with golden shoes, and from her was emanating the most splendid perfume, surpassing all aromas. She clothed the man with a purple robe, lifted him up to his brightest clearness, and took him with herself to Heaven." Therefore says SENIOR: " It is a living thing, which no more dies, but when used gives an eternal increase."

Plate IX. – THE FIFTH PARABLE

THE FIFTH PARABLE

The Philosophers give to this Art two bodies, namely: Sun and Moon, which are Earth and Water, they also call them Man and Wife, and they bring forth four children, two boys, which are heat and cold, and two girls, as moisture and dryness. These are the Four Elements, constituting the QUINTESSENCE, that is the proper white MAGNESIA, wherein there is nothing false. In conclusion SENIOR remarks: "When these five are gathered together, they form ONE substance, whereof is made the natural Stone, while AVICENA contends that: "if we may get at the Fifth, we shall have arrived at the end."

So let us understand this meaning better. The Philosophers take for example an Egg, for in this the four elements are joined together. The first or the shell is Earth, and the White is Water, but the skin between the shell and the White is Air, and separates the Earth from the Water; the Yolk is Fire, and it too is enveloped in a subtle skin, representing our subtle air, which is more warm and subtle, as it is nearer to the Fire, and separates the Fire from the Water. In the middle of the Yolk there is the Firm ELEMENT, out of which the young chicken bursts and grows. Thus we see in an egg all the elements combined with matter to form a source of perfect nature, just so as it is necessary in this noble art.

Plate X. – THE SIXTH PARABLE

THE SIXTH PARABLE

Rosinos relates of a vision he had of a man whose body was dead and yet beautiful and white like Salt. The Head had a fine Golden appearance, but was cut off the trunk, and so were all the limbs; next to him stood an ugly man of black and cruel countenance, with a bloodstained double-edged sword in his right hand, and he was the good man's murderer. In his left hand was a paper on which the following was written: "I have killed thee, that thou mayest receive a superabundant life, but thy head I will carefully hide, that the worldly wantons may not find thee, and destroy the earth, and the body I will bury, that it may putrefy, and grow and bear innumerable fruit."

Plate XI. – THE SEVENTH PARABLE

THE SEVENTH PARABLE

OVID the old Roman, wrote to the same end, when he mentioned an ancient Sage who desired to rejuvenate himself was told: he should allow himself to be cut to pieces and decoct to a perfect decoction, and then his limbs would reunite and again be renewed in plenty of strength.

Plate XII.-THE FOURTH TREATISE, FIRSTLY

THE FOURTH TREATISE

OF THE MEANS BY WHICH NATURE

ATTAINS HER ENDS

ARISTOTLE in the Book of Origins speaks thus: "SUN and Man create a Man, for the Sun's force and spirit give life, and the process has to be gone through seven times, by means of the Sun's heat." But as the Philosophers in their work have to aid Nature with Art, so have they also to govern the heat according to the Sun, so as to create the before-mentioned Stone, which as well has to undergo seven processes. For such a work requires **FIRSTLY**, a heat powerful enough to soften and melt these parts of the earth that have become thick, hard and baked, as mentioned by

SOCRATES when he says : that the holes and cracks of the earth will be opened to receive the influence of Fire and Water.

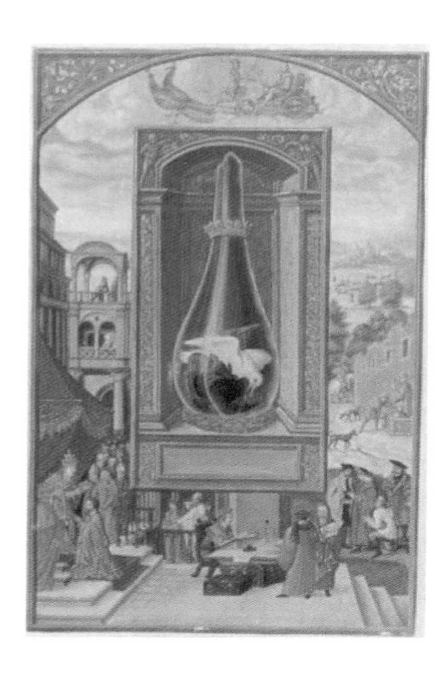

Plate XIII. – THE FOURTH TREATISE, SECONDLY

THE FOURTH TREATISE

SECONDLY: The Heat is necessary, because through its power the earth becomes freed from darkness and blessed with light instead. In regard to which SENIOR says: that heat turns every black thing white, and every white thing red. So, as water bleaches, fire gives off light, and also colour to the subtilized earth, which appears like a Ruby, through the tinging Spirit she receives from the force of the fire, thus causing SOCRATES to say: that a peculiar light shall be seen in the darkness.

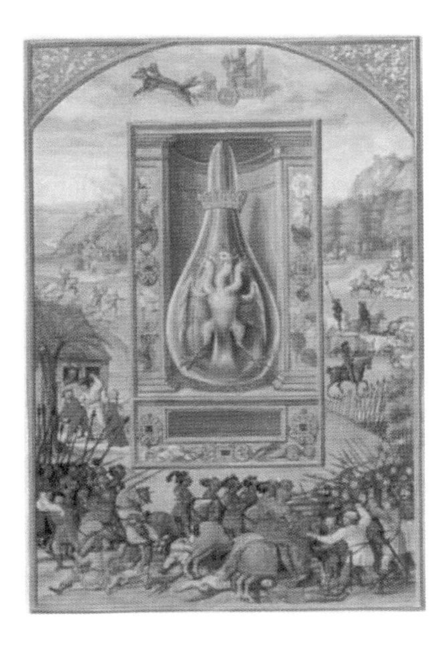

Plate XIV. – THE FOURTH TREATISE, THIRDLY

THE FOURTH TREATISE

THIRDLY: The Heat causes earthly things to be penetrated by a Spiritual Force, of which it is said in the TURBA: Spiritualize the bodies and make Volatile that which is Fixed. Of which RHAZES reminds in his LIGHT OP LIGHTS," as follows: "A heavy body cannot be made light without the help of a light body, nor can a light body be kept pressed down to the ground without the aid of a heavy body.

Plate XV.- THE FOURTH TREATISE, FOURTHLY

THE FOURTH TREATISE

FOURTHLY: The Heat cleanses that which is unclean. It throws off the mineral impurities and bad odours and nourishes the Elixir. In mention of which HERMES advises: Separate the gross from the Subtil, the earth from the fire. Whereof says ALPHIDIUS: The earth can be molten and becomes fire. Thereon says RHAZES: There are several Purifications preceding the perfect preparation, namely, Mundification and Separation.

Plate XVI.- THE FOURTH TREATISE, FIFTHLY

THE FOURTH TREATISE

FIFTHLY: The Heat works elevatingly, for by its force the spirits hidden in the Earth are raised up into the air, wherefore the Philosophers say, that whosoever can bring to light a hidden thing, is a Master of the Art.

The same is meant by MORIENUS, when he teaches that "he who can recreate the Soul is able to see colour, and also by ALPHIDIUS saying: "Hence it is that this Combat raises upwards, or else you shall not gain by it.

Plate XVII.- THE FOURTH TREATISE, SIXTHLY

THE FOURTH TREATISE

SIXTHLY: The Heat warms the cold earth, that while cold was half dead. Thereof says SOCRATES: When Heat penetrates, it makes subtle all earthly things, that are of service to the matter, but come to no final form while it is acting on the matter. The Philosophers conclude on the mentioned Heats in brief words, saying: Destil seven times and you have separated the destructable moisture and it takes place as in one destillation.

Plate XVIII.- THE FOURTH TREATISE, EIGHTHLY

THE FOURTH TREATISE

SIXTHLY: The Heat warms the cold earth, that while cold was half dead. Thereof says SOCRATES: When Heat penetrates, it makes subtle all earthly things, that are of service to the matter, but come to no final form while it is acting on the matter. The Philosophers conclude on the mentioned Heats in brief words, saying: Destil seven times and you have separated the destructable moisture and it takes place as in one destillation.

SEVENTHLY: Is the Force of the heat thus mixed with heat in the earth, that it has made light the collected parts and resolved them so as to surpass the other elements, and therefore this heat shall be modified with the Coldness of the Moon, "Extinguish the Fire of one thing with the Coldness of another" says CALID.

EIGHTHLY, AUCTOR DE TRIUM VERBORUM, the author of THE THREE WORDS gives in his writings a peculiar method to govern the HEAT or the FIRE, saying: "When the Sun is in Aries, he indicates the First Heat, or Grade of the Fire, which is weak because the heat is under the Rule of the Water, but when the Sun is in Leo, then it indicates the Second Grade, which is hotter because the great coldness of the Water being under the Rule of the Air. In the Sign of Saggitarius is the Third Grade, this being not of a burning heat, and under the Rule or Order of Rest and Pause.

Plate XIX. – THE FIFTH TREATISE, PART I, 1st CHAPTER

THE FIFTH TREATISE

ON THE MANIFOLD OPERATIONS OF THE WHOLE WORK IN FOUR CHAPTERS

THE FIRST CHAPTER

DISSOLUTION is the FIRST Operation which has to take place in the Art of ALCHEMY, for the order of Nature requires that the CORPUS, BODY, OR MATTER, be changed into WATER which is the much spoken of MERCURY. The LIVING SILVER dissolves the adjoined pure SULPHUR. This Dissolution is nothing but a killing of the moist with the dry, in fact a PUTREFACTION, and consequently turns the MATTER black.

Plate XX. – THE FIFTH TREATISE, PART I, 2nd CHAPTER

THE FIFTH TREATISE : PART I .

THE SECOND CHAPTER

The next is COAGULATION, which is turning the WATER again into the CORPUS or MATTER, meaning thereby that the SULPHUR, which before was dissolved by the LIVING SILVER, absorbs the same and draws it into itself. The Water that turned to Earth, which the Corpus has absorbed, necessarily shows other and manifold colours. For if the properties of an operating thing alter, so must the thing operated on alter. Because in the DISSOLUTION the LIVING SILVER is active, but in the COAGULATION it is passive, operated on. Wherefore is this Art compared to the play of children, who when they play, turn undermost that which before was uppermost.

Plate XXI. – THE FIFTH TREATISE, PART I, 3rd CHAPTER

THE FIFTH TREATISE

THE THIRD CHAPTER

The Third is SUBLIMATION, distilling the before-mentioned moisture of the earth, for if the water is reduced into the earth, it is evaporated into the lightness of the air, and rises above the earth, as an oblong cloudlet, like an egg, and this is the Spirit of the FIFTH ESSENCE, which is called the TINCTURE, ANIMA, FERMENTUM, or the OIL, and which is the very next matter to the STONE OF THE PHILOSOPHERS.

Plate XXII. – THE FIFTH TREATISE, PART I, 4th CHAPTER

THE FIFTH TREATISE

THE FOURTH CHAPTER

The fourth Chapter sheweth the last or fourth thing belonging to this water which has been separated from the earth, be again joined to the earth. The one thing must be done with the other, if the Stone is to be made perfect.

The reason why all natural things are put together in body is, that there may be a united composition.

In these last four Chapters is all contained where with the Philosophers have filled the whole world with innumerable books.

Made in the USA
San Bernardino, CA
13 August 2013